Climates of the World:
Identifying and Comparing
Barbara M. Linde
AR B.L.: 6.0
Points: 0.5

Climates of the World

Identifying and Comparing Mean, Median, and Mode

Barbara Linde

PowerMath™

The Rosen Publishing Group's

PowerKids Press™

New York

Published in 2005 by The Rosen Publishing Group, Inc.
29 East 21st Street, New York, NY 10010

Book Design: Haley Wilson

Photo Credits: Cover, pp. 17 (wheat), 22, 24 (rabbit) © PhotoDisc; p. 5 © Michael S. Yamashita/Corbis;
p. 7 courtesy of Environment and Natural Resources Service FAO of the UN, Rome; p. 8 (parrots) © Alan
Schein Photography/Corbis; p. 8 (butterfly) © Danny Lehman/Corbis; p. 9 © Galen Rowell/Corbis;
pp. 10 (gazelle), 15 (camel), 21, 23 © Royalty Free/Corbis; p. 10 (zebras, giraffes) © Artville;
pp. 10 (lions), 29 © Digital Vision; p. 11 © Brian A. Vikander/Corbis; p. 13 © Tiziana and Gianni
Baldizzone/Corbis; p. 15 © Photowood Inc./Corbis; p. 17 © David Stoecklein/Corbis; p. 19 provided
by Terra Satellite, EOS, NASA; p. 20 © Layne Kennedy/Corbis; pp. 24 (bear), 26 (walrus, artic fox), 27
(caribou), 28 © Digital Stock; p. 25 © William Manning/Corbis; p. 26 (polar bears) © Randy Wells/Corbis;
p. 27 © Paul A. Souders/Corbis.

Library of Congress Cataloging-in-Publication Data

Linde, Barbara M.
  Climates of the world : identifying and comparing mean, median, and mode / Barbara Linde.
      p. cm. — (PowerMath)
  Includes index.
  ISBN 1-4042-2932-9 (lib. bdg.)
  ISBN 1-4042-5125-1 (pbk.)
  6-pack ISBN: 1-4042-5126-X
  1. Average—Juvenile literature. 2. Climatology—Juvenile literature. I. Title. II. Series.
  QA115.L682 2005
  519.5'33—dc22
                        2004003213

Manufactured in the United States of America

Contents

What Is Climate?

Earth has many different kinds of climates, ranging from deserts to rain forests to **tundras**. To better understand Earth's vastly different places, we need to look at the differences in Earth's climates. Sometimes people confuse climate and weather. Weather is the day-by-day condition of the atmosphere. Climate is the long-term weather of a place over many years. You may live in a warm climate, but there may be days of sunny, rainy, hot, and cool weather throughout the year. Climate includes average temperature, **precipitation**, **humidity**, and wind conditions.

The latitude of a location influences its climate. The closer a place is to the equator, the hotter the climate is. Places at the same latitude on different continents have similar climates. Elevation, or height above sea level, also influences climate. Wind and ocean currents affect climate, too.

When we talk about climates, we can use the math terms "mean," "median," and "mode." If we take the total amount of something and divide it by the number of figures we added together to get that amount, the result is called the mean. Another word for mean is "average." If we put a group of numbers in increasing order, the number in the middle is called the median. The number that occurs the most often in a group of numbers is called the mode.

We can better understand mean, median, and mode by taking a look at the daily temperatures in New York City from February 1–7, 2003.

Date	Feb 1	Feb 2	Feb 3	Feb 4	Feb 5	Feb 6	Feb 7
Temp	37°F	45°F	50°F	48°F	37°F	32°F	33°F

$$\text{mean} = \frac{\text{sum of numbers}}{\text{number of numbers}} = \frac{37 + 45 + 50 + 48 + 37 + 32 + 33}{7} = \frac{282}{7} = 40.29$$

When we round to the nearest tenth, the **mean** temperature for the week is **40.3°F**.

To find the median, arrange the numbers in increasing order: 32, 33, 37, 37, 45, 48, 50. The median is the number in the middle, or 37. The **median** temperature is **37°F**.

To find the mode, see how many times each number repeats: 37, 45, 50, 48, 37, 32, 33. The only number that repeats is 37, so the **mode** is **37°F**.

New York City

In 1918, a German man named Vladimir Koeppen introduced a system that grouped Earth's climates into 5 main climate zones—tropical, dry, temperate (or mild), cold, and polar. These climate zones were based on monthly and annual averages of temperature and precipitation. Koeppen believed that climate zones matched up with natural patterns of soil and the kinds of plant life the soil supported. Koeppen's system is still widely used by scientists today.

Koeppen was a climatologist (kly-muh-TAH-luh-jist). A climatologist is a scientist who studies Earth's climates. Climatologists keep records of temperatures and amounts of rainfall and snowfall over a period of time. They track the number of sunny, cloudy, or foggy days. They also record the number of severe storms in different locations. **Hurricanes, tornadoes,** and blizzards are all severe storms that can affect the climate of a place. Climatologists use all of this information to better understand climate and to chart changes in climate that affect Earth. Finding the mean, median, and mode is an important part of this process.

We can look at the amount of January snowfall in inches in Milwaukee, Wisconsin, from 1999 to 2003 to find the mean amount of snowfall for the month of January.

1999	2000	2001	2002	2003
39.0	15.2	1.3	13.1	3.7

$$\text{mean} = \frac{39.0 + 15.2 + 1.3 + 13.1 + 3.7}{5} = \frac{72.3}{5} = 14.46$$

When we round to the nearest tenth of an inch, the mean January snowfall in Milwaukee is about **14.5 inches.**

The 5 main climate zones identified in Koeppen's system are: tropical, dry, temperate (or mild), cold, and polar. Koeppen used different colors to show the zones. At a glance, you can tell which areas have similar climates. The map shows how climates change from the equator to the North and South Poles. Where might you go if you wanted to take a vacation in a warm, sunny place?

Koeppen's Climate Classification System

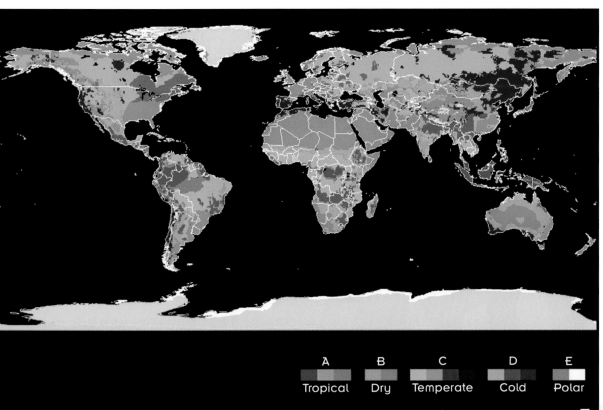

A	B	C	D	E
Tropical	Dry	Temperate	Cold	Polar

Tropical Climates

According to Koeppen's system, there are 3 tropical climates: the tropical wet, tropical wet and dry, and tropical **monsoon**.

The tropical wet climate occurs in Earth's rain forests close to the equator. The mean monthly temperature of a rain forest is around 77°F, and the average yearly rainfall is more than 100 inches. In fact, 260 inches of rain might fall in a single year! The heavy rain makes the air hot and humid.

The heavy rain also makes the trees in the rain forest grow tall. Some can reach heights of over 160 feet. These trees create a **canopy** over the forest. This thick layer of leaves blocks much of the sunlight from the smaller trees in the rain forest's **understory**. The forest floor also gets little sun, making it a perfect place for ferns, mosses, and other shade-loving plants to grow.

Butterflies, mosquitoes, and hundreds of other insect species live in the tropical rain forests. Brightly colored parrots, frogs, and snakes also make their home there, as do tigers, elephants, and many other animals.

Manaus, Brazil, is in the Amazon rain forest in South America. The chart below can help us find the median high temperature there for the period from December 28, 2003, through January 3, 2004.

Date	Dec 28	Dec 29	Dec 30	Dec 31	Jan 1	Jan 2	Jan 3
Temp	89°F	91°F	89°F	89°F	89°F	88°F	87°F

First, arrange the temperatures in increasing order:
87°F, 88°F, 89°F, 89°F, 89°F, 89°F, 91°F.

The temperature in the middle, 89°F, is the median.

Amazon rain forest

The **savanna** region has a wet and dry tropical climate. It is also called a tropical grassland. Savannas are located on the borders of tropical rain forests and are mostly made up of tall grasses that can grow from 3 to 10 feet high. There are also some bushes and shrubs, but not many trees.

The savanna has distinct wet and dry seasons. There is usually no rain at all during the dry season, and the temperature stays between 68°F and 78°F. During the wet season, there is usually between 10 and 30 inches of rainfall, and temperatures range from 78°F to 86°F.

Plant eaters like the gazelle, giraffe, and zebra all live on the Serengeti Plain and eat the tall grasses and shrubs that grow there. Lions also roam the plain and hunt these animals for food

The Serengeti Plain in Tanzania (tan-zuh-NEE-uh), Africa, is one of the world's largest savannas. Much of it is now a national park. Below are the mean monthly temperatures of the Serengeti Plain. We can use this information to find the mode—the mean monthly temperature that occurs most often.

Jan	Feb	Mar	Apr	May	June	July	Aug	Sept	Oct	Nov	Dec
70°F	70°F	70°F	70°F	66°F	63°F	63°F	64°F	66°F	68°F	70°F	70°F

What mean monthly temperature occurs most often? That temperature is the mode.

Serengeti National Park

Take along a raincoat, boots, and an umbrella if you ever visit a tropical monsoon area! A monsoon is a wind that blows in one direction for part of the year, then reverses and blows in the other direction for the rest of the year. In the summer, moist air from the warm ocean blows over the land. This moist air brings very heavy rain. In the winter, the land cools down and dry air blows in the opposite direction, from the land to the sea.

Monsoons occur mostly in southern and eastern Asia, bringing heavy rains to countries like India, Thailand, Burma, and Bangladesh. Monsoons also blow over South America, the coasts of northern Australia, parts of Africa, and the southwestern United States. In most of these areas, the rainy season starts around April or May and ends by October. The amount of yearly rainfall varies from place to place in monsoon areas around the world, but usually ranges from 30 to 150 inches.

The strength of a monsoon can affect an area's economy. Very strong monsoon winds can destroy livestock and crops. The greatest amount of rainfall ever recorded during a monsoon was in Cherrapunji, India, which received 893 inches in 1 year. That's more than 74 feet of rain!

This photograph shows a rural village in Orissa, a state in eastern India, during monsoon season. Though monsoons can be destructive, they also bring much needed rain to areas that often experience long periods with little rainfall.

Below are the average monthly rainfall totals in inches for Cherrapunji, India, over a 20-year period. We can use these monthly averages to find the mean amount of monthly rainfall over this 20-year period. All measurements are rounded to the nearest inch.

Jan	Feb	Mar	Apr	May	June	July	Aug	Sept	Oct	Nov	Dec
0	2	9	37	47	90	128	69	53	21	3	1

$$\text{mean} = \frac{\text{sum of numbers}}{\text{number of numbers}} = \frac{460}{12} = 38.33$$

The mean amount of monthly rainfall for this 20-year period was 38.33 inches, or about **38 inches** per month.

Dry Climates

About 30% of Earth has a dry climate. There are 2 kinds of dry climates: the desert climate and the **semiarid**, or **steppe**, climate.

The desert climate is arid, or very dry. Hot deserts are warm in the fall and spring and extremely hot in the summer. Winters can be cool. The hot, dry desert gets less than 10 inches of rainfall a year. Daytime temperatures can go above 120°F. At night, temperatures can drop below 40°F.

Cold deserts can have temperatures as low as −40°F. The Arctic and Antarctic regions both have cold deserts. Even though there is a lot of ice, the air is too cold and dry to make snow.

Tucson, Arizona, is in the Sonoran Desert. This desert covers about 120,000 square miles, including the southwestern United States and part of Mexico. The chart below shows Tucson's monthly rainfall in inches for a year. Use it to find the mode—the number that repeats most often.

Jan	Feb	Mar	Apr	May	June	July	Aug	Sept	Oct	Nov	Dec
0.99	0.88	0.81	0.28	0.24	0.24	2.07	2.30	1.45	1.21	0.67	1.03

Only 0.24 repeats, so **0.24 inches** is the mode.

The Sahara in Africa is the world's largest hot desert. The hottest temperature ever recorded happened there on September 13, 1922, in a country called Libya in northern Africa. On that day, the temperature reached 136°F!

Sahara desert

Semiarid, or steppe, climates are often located between a desert and an area that receives more rainfall. The steppe climate is not usually close to large water sources like oceans. Mountains sometimes keep moist ocean air out, and the steppe only gets between 10 and 20 inches of rain a year. There may be as many as 10 years with rain, followed by years of **drought**.

Since the air is dry, there aren't many clouds. This means the warm air from the ground escapes into the atmosphere more easily and the land can get very cold. In the winter, the temperature can drop as low as −40°F. Summer temperatures can reach 120°F.

A steppe climate region is made up mostly of grassland, small, low plants, and few trees. Alice Springs, Australia, is located near the Great Australian Desert. Alice Springs is in the steppe climate zone.

The chart shows average monthly temperatures for Alice Springs. Find the mean temperature for March, April, and May.

Jan	Feb	Mar	Apr	May	June	July	Aug	Sept	Oct	Nov	Dec
82°F	82°F	77°F	68°F	59°F	54°F	54°F	57°F	64°F	73°F	79°F	80°F

$$\text{mean} = \frac{\text{sum of numbers}}{\text{number of numbers}} = \frac{77 + 68 + 59}{3} = \frac{204}{3} = 68$$

The mean temperature for March, April, and May in Alice Springs is **68°F.**

Steppe climates are found throughout the world. In North America, the steppe covers much of the Great Plains from northern New Mexico northward into Canada. Much of the land in this large area is used to grow crops like wheat and to graze livestock. The photograph below shows an area of the Great Plains in Nebraska in the United States.

Temperate Climates

The word "temperate" means "mild." In the middle latitudes, the climate is moist. Winters are mild, and there are seasonal changes in temperature. There are several climate subgroups within this main group: the humid subtropical climate, the Mediterranean climate, and the marine west coast climate.

Humid subtropical climates are found in the eastern parts of the continents. The term "subtropical" refers to an area that borders a tropical climate zone. In the humid subtropical zone, summer temperatures usually range from 75°F to 90°F. Winter temperatures range from 30°F to 50°F. There is rain all year, but most of the rain falls during the summer. There might be some snow or frost in the winter.

Humid subtropical climates experience many thunderstorms, hurricanes, and **typhoons**. The southeastern United States has a humid subtropical climate and experiences hurricanes, usually during the late summer months of August and September. A hurricane's winds swirl around the hurricane's "eye," a calm zone located in the hurricane's center that is usually between 10 and 20 miles wide. During a hurricane, wind speeds outside of the eye can measure up to 200 miles per hour!

Hurricanes are storms that start over warm oceans. They bring huge waves, strong winds, and heavy rainfall when they reach land. Powerful hurricanes can destroy buildings and often kill people as well. In this image, we can see hurricane winds swirling around the hurricane's center, or "eye," in the Atlantic Ocean.

The chart below shows the number of Atlantic hurricanes that happened each year from 1999 to 2003. We can use it to find the median and mode.

Year	1999	2000	2001	2002	2003
# of Hurricanes	8	8	9	4	7

To find the median, arrange the numbers in increasing order: 4, 7, 8, 8, 9. The median is the number in the middle, or 8, so the median number of hurricanes is 8.

To find the mode, see how many times each number repeats. Since 8 is the only number that repeats, 8 is also the mode.

The Mediterranean climate is found on the western coasts of most of the continents (except Antarctica) and does not go far inland. Winters are mild, with temperatures around 50°F and about 20 inches of rain. Summers are sunny and warm with temperatures from around 75°F to 85°F and very little rain. Summer temperatures can go as high as 100°F.

Marine west coast climate zones have mountain ranges that run north to south near the coasts. Wind from the ocean blows over the land, keeping the air humid and the temperatures mild. The mountains block the winds from going inland. Most areas with a marine west coast climate get between 30 and 50 inches of rain a year. Some areas get up to 150 inches! In these areas, there are many cloudy days with fog and drizzle. Summer temperatures range from 60°F to 70°F. Winter temperatures range from 35°F to 45°F.

Trees called redwoods grow from San Francisco, California, to southern Oregon in a marine west coast climate. Redwoods are also called giant sequoias. They can grow to be over 300 feet tall and can live as long as 2,000 years. The photograph on the left shows a fallen giant redwood tree that is thought to be about 1,000 years old!

Portland, Oregon, has a marine west coast climate. The chart below shows the high temperatures in Portland on June 1 from 1996 through 2003. What is the mean high temperature for June 1?

1996	1997	1998	1999	2000	2001	2002	2003
79°F	66°F	73°F	58°F	78°F	65°F	74°F	72°F

$$\text{ean} = \frac{\text{sum of numbers}}{\text{number of numbers}} = \frac{79 + 66 + 73 + 58 + 78 + 65 + 74 + 72}{8} = \frac{565}{8} = 70.63$$

hen we round to the nearest tenth of a degree, the mean high temperature
r June 1 is **70.6°F**.

Cold Climates

There are also several types of cold climate. In these areas, the mean temperature of the coldest month is below freezing, or 32°F. Winters are long and very cold. There are several subgroups of the cold climate. This chapter will describe the humid continental climate and the subarctic climate.

The word "continental" means that the climate zone is located in the interior of the continent. Northeastern and central North America, Europe, and Asia all have humid continental climates. These areas have 4 separate seasons—winter, spring, summer, and fall. Temperatures change greatly from season to season and sometimes from day to day. The average temperature during hot months is usually in the mid-70s°F. The annual precipitation ranges from 20 to 50 inches. It rains in the summer, spring, and fall, and snows in the winter. Areas with humid continental climates have good farming because of the steady precipitation they receive.

Most deciduous forests are found in the humid continental climate. The word "deciduous" means "falling off." Deciduous forests have trees whose leaves fall off in autumn. In spring, new leaves grow.

Duluth, Minnesota, has a humid continental climate. Below are the high temperatures for the week of January 1–7, 2002.

Date	Jan 1	Jan 2	Jan 3	Jan 4	Jan 5	Jan 6	Jan 7
Temp	18°F	18°F	23°F	27°F	30°F	28°F	25°F

What is the mean high temperature for the week? You can add the temperatures and divide by 7 to find your answer.

The subarctic climate is north of the humid continental climate, and borders the Arctic—or polar—climate zone. The subarctic climate is only found in the northern hemisphere because there are no large areas of land at the same latitude in the southern hemisphere.

The subarctic climate is a cold, snowy climate with a winter that lasts for 6 or 7 months. Winter temperatures range from –65°F to 30°F. Even the summer is cool, with temperatures ranging from 50°F to around 70°F. The subarctic climate has the greatest differences in temperatures of any climate on Earth. Areas with a subarctic climate get about 5 to 20 inches of precipitation per year, mostly in the form of summer rain. In the winter, it is too cold and dry for it to snow.

The subarctic climate zone has a large evergreen fore where animals such as bears, rabbits, reindeer, ar beavers make their homes

Below are the average monthly temperatures for Fort Smith, Canada, which has a subarctic climate. Can you find the median monthly temperature?

Jan	Feb	Mar	Apr	May	June	July	Aug	Sept	Oct	Nov	Dec
-14°F	−6°F	7°F	30°F	47°F	57°F	62°F	58°F	46°F	33°F	10°F	−6°F

rrange the numbers in increasing order. Since negative numbers are lower mperatures, start with them:

14, −6, −6, 7, 10, 30, 33, 46, 47, 57, 58, 62.

ne median is the number in the middle. Since there is an even number of numbers, e add the two middle numbers, then divide by 2:

0 + 33 = 63   63 ÷ 2 = 31.5   The median monthly temperature is **31.5°F**.

Polar Climates

The polar climate zones cover the areas around the North and South Poles, and include the tundra and ice cap climates.

The tundra climate is between the subarctic climate and the ice cap climate. It is mostly found near the Arctic Circle. A small tundra climate zone is also found along the coast of Antarctica. The average winter temperature is between −20°F and −30°F. During the winter, the wind constantly howls and it is almost always dark. The tundra climate gets only about 6 to 10 inches of precipitation per year. The summer lasts only about 10 weeks, and summer temperatures never go above 50°F.

The tundra zone is treeless. The ground has a permanently frozen layer called **permafrost**. The surface of the permafrost thaws during the short summer, allowing low plants and flowers to grow.

Despite the cold climate, the tundra is home to wildlife such as polar bears, walruses, seals, caribou, wolves, arctic foxes, arctic hares, squirrels, and porcupines.

Barrow, Alaska, has a tundra climate. Below are the average monthly wind speeds for Barrow, shown in miles per hour.

Jan	Feb	Mar	Apr	May	June	July	Aug	Sept	Oct	Nov	Dec
12	11	11	11	12	11	11	12	13	13	13	11

Can you find the mode?

Remember, the mode is the number that occurs the most often.

The ice cap climate is found in Greenland near the North Pole and in Antarctica around the South Pole. This is the coldest climate on Earth—no month has an average temperature above 32°F. Usually no more than 5 inches of precipitation fall during the year, and the ice in these places never melts. This climate is sometimes called a "polar desert."

About 55,000 people live in Greenland. Whales, seals, polar bears, musk oxen, and reindeer also live there. Antarctica does not have any native residents. However, depending on the time of year, about 1,000 to 4,000 people live there to research the area. Whales, seals, and penguins live in the waters around Antarctica.

Antarctica has the largest area of ice cap climate. It is the coldest, driest, and windiest of all climate zones. On July 21, 1983, at Vostok Station in Antarctica, the temperature went down to −128.6°F! That is the coldest day ever recorded.

Antarctica, shown at the right, has an area of about 5,400,000 square miles. That is larger than the continents of Australia and Europe!

Below are high temperatures for a December day at 9 places in Greenland. Find the median high temperature.

Aputiteeq	16°F
Ikermiit	11°F
Kulusuk	15°F
Nuuk	19°F
Pituffik	9°F
Qaanaaq	1°F
Qaarsut	12°F
Qaqortoq	21°F
Tasiilaq	17°F

Arrange the numbers in increasing order:
1, 9, 11, 12, 15, 16, 17, 19, 21.

The median is the number in the middle: 15.

The median temperature for these places is 15°F.

Using Mean, Median, and Mode

You can record the daily temperatures and amounts of precipitation where you live, then use mean, median, and mode to describe your climate.

We now know that the mean is the sum of all the numbers divided by the number of numbers. Find the mean monthly rainfall in Hilo, Hawaii, if the mean annual rainfall is 126 inches.

mean = $\dfrac{\text{sum of numbers}}{\text{number of numbers}}$ = $\dfrac{126}{12}$ = 10.5 The mean monthly rainfall is **10.5 inche**

We know that the median is the number in the middle when numbers are put in increasing order. Below are the monthly rainfall totals in inches for St. Louis, Missouri, in increasing order. Find the median. Remember, if there is an even amount of numbers, add the 2 middle numbers, then divide by 2.

2.14	2.28	2.76	2.86	2.96	2.98	3.60	3.69	3.71	3.76	3.90	4.11

2.98 + 3.60 = 6.58   6.58 ÷ 2 = 3.29
When we round to the nearest tenth of an inch, the median monthly rainfall is 3.3 inches.

We know that the mode is the number that occurs most often. Here are the monthly rainfall totals in inches for San Diego, California, for 1 year.

Jan	Feb	Mar	Apr	May	June	July	Aug	Sept	Oct	Nov	Dec
2.28	2.04	2.26	0.75	0.20	0.09	0.03	0.09	0.21	0.44	1.07	1.31

Can you find the mode?

Glossary

canopy (KA-nuh-pea) The highest layer of branches in a forest.

drought (DROWT) A long period of dry weather.

humidity (hyoo-MIH-duh-tee) The amount of moisture in the air.

hurricane (HUR-uh-kayn) A severe storm with very strong, dangerous winds and usually very heavy rain that occurs in the Atlantic Ocean.

monsoon (mahn-SOON) A wind that blows in one direction for part of the year, then reverses and blows in the other direction for the rest of the year. Monsoons usually bring heavy rain.

permafrost (PURH-muh-frost) A permanently frozen layer below the surface of the ground in polar climates.

precipitation (prih-sih-puh-TAY-shun) Water that falls to Earth as rain, snow, sleet, or hail.

savanna (suh-VA-nuh) A grassy plain with few or no trees.

semiarid (seh-mee-A-ruhd) Having little rainfall, usually between 10 and 20 inches per year.

steppe (STEP) A vast, treeless plain.

tornado (tohr-NAY-doh) A very strong windstorm with winds as high as 200 miles per hour. Tornadoes form twisting columns of air that extend down from clouds and move across the land.

tundra (TUHN-druh) A vast, flat, treeless plain in the arctic regions.

typhoon (ty-FOON) A storm with strong wind and very heavy rain that occurs near eastern Asian countries.

understory (UHN-duhr-stohr-ee) The layer of trees and shrubs between the forest canopy and the ground.

Index

DATE DUE

~~FEB 0 5~~			
FEB 2 2			

FOLLETT